EXPLORING
MARS

By Nick Christopher

KidHaven
PUBLISHING

Published in 2018 by
KidHaven Publishing, an Imprint of Greenhaven Publishing, LLC
353 3rd Avenue
Suite 255
New York, NY 10010

Designer: Deanna Paternostro
Editor: Vanessa Oswald

Photo credits: Cover, back cover, pp. 5, 21 Vadim Sadovski/Shutterstock.com; p. 7 MSSA/Shutterstock.com; p. 9 Aaron Rutten/Shutterstock.com; pp. 10–11 Jurik Peter/Shutterstock.com; p. 13 KEES VEENENBOS/SCIENCE PHOTO LIBRARY/Getty Images; p. 15 NASA/JPL/ARIZONA STATE UNIVERSITY/SCIENCE PHOTO LIBRARY/Getty Images; p. 17 Image copyrights Moyan Brenn; p. 19 SergeyDV/Shutterstock.com.

Cataloging-in-Publication Data

Names: Christopher, Nick.
Title: Exploring Mars / Nick Christopher.
Description: New York : KidHaven Publishing, 2018. | Series: Journey through our solar system | Includes index.
Identifiers: ISBN 9781534522862 (pbk.) | 9781534522701 (library bound) | ISBN 9781534522558 (6 pack) | ISBN 9781534522626 (ebook)
Subjects: LCSH: Mars (Planet)–Juvenile literature.
Classification: LCC QB641.C575 2018 | DDC 523.43–dc23
Printed in the United States of America

CPSIA compliance information: Batch #BS17KL: For further information contact Greenhaven Publishing LLC, New York, New York at 1-844-317-7404.

Please visit our website, www.greenhavenpublishing.com. For a free color catalog of all our high-quality books, call toll free 1-844-317-7404 or fax 1-844-317-7405.

CONTENTS

THE RED PLANET

Mars is known as the "Red Planet" because of its red color as seen from Earth. It's the fourth planet from the sun in our **solar system**.

Mars isn't just red. Other colors on its surface include tan, brown, gold, and specks of green.

A DAY AND YEAR ON MARS

A day on Mars is about 40 minutes longer than a day on Earth. A year on Mars is about 687 Earth days long. This is how long it takes for Mars to **orbit** the sun.

Mars

This picture shows how close Mars is to the sun, Earth, and other planets.

MARS'S MOONS

Mars has two small moons named Phobos (FOH-bohs) and Deimos (DAY-mohs). Phobos is bigger than Deimos. These moons are different than most moons because they aren't round.

Astronomer Asaph Hall discovered Phobos and Deimos, shown here, in 1877.

DESERT PLANET

Mars has been called a desert planet because it's covered with sand. The iron in the sand gives the planet its red color.

Mars often has dust storms, which can spread across the whole planet!

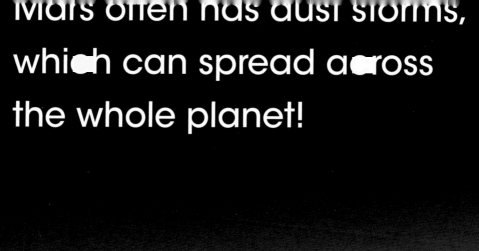

Mars has many sand dunes, such as the ones shown here.

Mars's **crust** has many mountains and valleys. It also has **volcanoes**, including Olympus Mons. This volcano is three times taller than Mount Everest, which is Earth's tallest mountain.

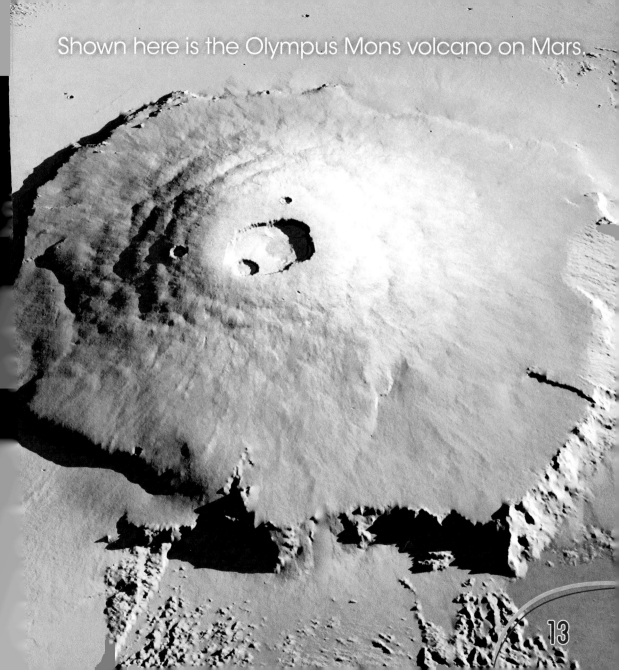

Shown here is the Olympus Mons volcano on Mars.

13

There are many **canyons** on Mars, including what's believed to be the deepest canyon in the solar system. It's about four times deeper than the Grand Canyon, which is Earth's deepest canyon.

canyons

These canyons are part
of the Valles Marineris,
which is the largest
canyon system on Mars.

SEARCHING FOR LIFE ON MARS

In 2015, scientists discovered flowing water on Mars. This proves there was possibly life on the planet in the past. Mars also has ice at its **poles** and below the ground's surface.

Flowing water on Mars
could also help support
future life on the planet!

Many **probes** have been sent to Mars to study the planet to help people learn more about it. These probes help scientists see what life is really like on Mars.

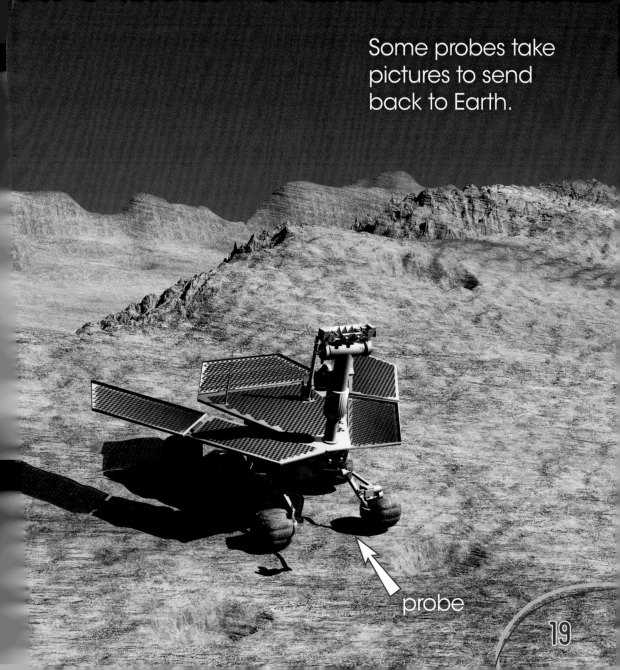

Some probes take
pictures to send
back to Earth.

probe

TRAVELING TO MARS

Astronauts are planning to go to Mars in the 2030s. Before this journey to Mars, they will be sent to an asteroid. Some scientists think people could even live on Mars someday!

It would take about seven months to get from Earth to Mars!

GLOSSARY

astronomer: A scientist who studies different parts of the solar system.

canyon: A deep, narrow valley with tall cliffs on the sides.

crust: The outer shell of a planet.

orbit: To travel in a circle or oval around something.

pole: The northernmost or southernmost point of a planet.

probe: A vehicle built to send information about an object in space back to Earth.

solar system: The sun and all the space objects that orbit it, including the planets and their moons.

volcano: An opening in a planet's surface through which hot, liquid rock sometimes flows.

FOR MORE INFORMATION

Websites

NASA: Mars
www.nasa.gov/content/journey-to-mars-overview
Readers can use this website to learn about Mars and look at pictures of the planet.

National Geographic Kids: Mission to Mars
kids.nationalgeographic.com/explore/space/ mission-to-mars/#mars-planet.jpg
This website provides useful facts about Mars.

Books

Bloom, J.P. *Mars.* Minneapolis, MN: ABDO Kids, 2015.

Carney, Elizabeth. *Mars.* Washington, DC: National Geographic Children's Books, 2014.

Owen, Ruth. *Mars.* New York, NY: Ruby Tuesday Books, 2014.

INDEX